WET PLATES AND DRY GULCHES

by

Gilbert L. Campbell

FILTER PRESS

Palmer Lake, Colorado

1970

WILD & WOOLLY WEST SERIES

1.	Choda	Thirty Pound Rails
2.	Clemens	Celebrated Jumping Frog
3.	Banks	Uncle Jim's Pancakes
4.	Service	Yukon Poems
5.	Cushing	My Adventures in Zuni
6.	Englert	Oliver Perry Wiggins
7.	Matthews	Navajo Weavers & Silversmiths
8.	Campbell	Wet Plates & Dry Gulches
9.	Banks	Alferd Packer's Cookbook
10.	Faulk	Simple Methods Mining Gold
11.	Rusho	Powell's Canyon Voyage
12.	Hinckley	Transcontinental Rails
13.	Young	The Grand Canyon
14.	Gehm	Nevada's Yesterdays **

** In preparation

SBN 910584-11-7

Printed in The
United States of
America

for Lollie

THE ROAD TO THE MINES

Preface

This little book is not intended to be a guide to the old Ghost towns or old mining camps. Rather, it is a collection of the camps as they were during their early years.

Although photography was a thriving art during the last part of the 19th century, for much of this period mass reproduction of pictures could be made only through the skill and painstaking work of the engravers who transferred each picture to carved wood blocks. Skilled engravers carved into the smooth end-grain surface of a wood block all detail that was to be printed. Heavy lines of uncut wood made the dark areas, while shading toward the light areas was made by carving hundreds of fine, almost parallel lines across the printing surface. Details were carved into the wood, and were shaded according to their nature. The individual interpretations rendered by the wood engravers are a part of the charm of these old pictures, and have remained as memories of the romance of the Old West. Their styles influenced the design of type faces in vogue at the same time, many of which were ornate, completely Victorian, and charming. Recently there has been a return of popularity of some of these old type faces which now serve in advertising.

The engravers were victims of their own progress. Their art had created a demand for more illustrations. With this demand came the need for a simpler method of reproducing photographs, and for eliminating the many hours of hard work needed to produce a small picture. After 1885 photographic halftone illustrations became increasingly popular. They didn't represent the subjects nearly as well from a technical standpoint, but they were PROGRESS, and they did improve as experience was gained. By the turn of the century most wood engravings were remnants of previous publications, or were used for special effects in catalogs or other pictorial works. Today there are few true commercial wood engravers practising the art. Most of their work is used for sophisticated advertising requiring special pictorial treatment.

Wood engraving was not a primitive process. The wood blocks were printed in proof presses so corrections or changes could be identified and made. Portions to be changed were cut out, and replaced with patches of fresh wood, which was carved to show the new detail, matching the old perfectly. (We still haven't learned to do this effectively with halftones!) When a picture was approved in its final form, an electrotype was made by electroplating a wax impression of the wood surface, making a perfect metal replica which would stand long press runs. Some of the early books had large printings, with editions of 100,000 copies or more before the turn of the century.

Many of these books played important roles in the development of the West. They were sold to readers in the great population centers of the East. Some were given away in the hope of attracting speculators or investors. Almost every book helped bring people or money to the growing West, thus speeding its development. Probably the four major factors in the growth of the West were publishing, railroads, photography, and the dream of finding a bonanza.

In this little book you will see some of the pictures that may have helped your Grandpa decide to come west to make his fortune. Or perhaps they persuaded him to lose his shirt in a poor investment. They are the same pictures that also probably convinced Grandma that the West was a horrible, wild, uncouth sort of place. The pictures split families, built fortunes, and caused hardships.

Because the California gold rushes began before the surge in publishing, most of the pictures here will show mining in the Rocky Mountains. The camps are different, but the methods of mining were much the same, and the ways of the lonely men were almost identical.

To be successful the early mining camps required a great deal of water. A dry gulch was less profitable, but might be a handy place for a bad guy to hole up. If you ventured through carrying your poke to the nearest saloon, or even to the Wells Fargo agent, this bad guy just might take a pot shot at you, and take your poke into town himself, claiming that he had dug it on his dry gulch claim. Hence, the term dry gulch came to be used for ambushing and robbing stages, express carriers, and others. These varmints didn't just lie in wait of rich passing miners, or the express coach, but would sometimes sneak up on a lonely miner and take over his claim. Claim registrations weren't the most accurate in the world, so if a man moved on another would take his place. If he died his claim did not go to his widow, so anyone could take it over. Dry gulchers often took over claims without making completely honest reports on the last illness or the departure of previous owners. Later competition got pretty tough for the bad guys, so many of them also met untimely ends. A few are shown here and there in this little book. Some were hung, some were banished, and some got rich exploiting likker, gambling, and girls.

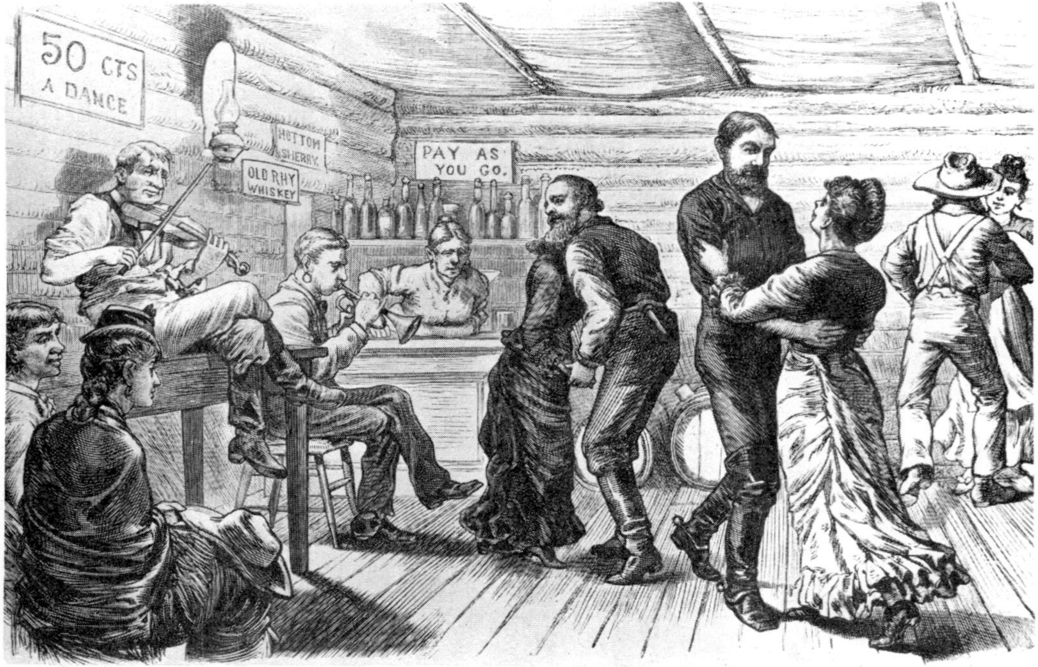

A DANCE HOUSE IN THE "WONDERFUL MINING TOWN OF LEADVILLE", 1879

What is now the United States was originally explored because of gold. Gold had sent the Spanish explorers to the New World where they conquered native kings, converted the Indians, and, incidently, shipped home riches for the Spanish Kings. Even today some of the sunken treasure ships are being salvaged, producing gold and silver in satisfactory quantities.

The Spanish Conquistadores came to the New World seeking gold, silver, pearls, and gems. This first great gold rush of history promised them the wealth to return to Spain where they could buy land and live at ease. Many did not return, of course, but helped found the European civilization in Spanish America.

The Indians had mined gold because they could recognize it, and could work it easily. Silver was mined by Indians only around Taxco, Mexico, as native silver, and was used as a natural silver-copper alloy. The gems taken included pearls from Baja California, and Panama, emeralds from Columbia, and turquoise, opals, and other precious stones from Mexico and Central America.

At first the precious metal and gems were plundered from the supplies accumulated by the Indians. Early placer mining began in 1515 in Hispaniola, then spread to Cuba, Puerto Rico, Mexico, Peru, and Venezuela. By 1531 Silver was being mined in Michoacan, Mexico, where the first American mining laws were introduced in 1532. Ore dressing and metallurgical processes used were far more efficient than most used in the United States until 1880 or so. German metallurgists introduced the best European processes to Mexico in the middle of the sixteenth century. For over 300 years Mexican Silver production averaged over $12.00 per minute. The 16th century metallurgists in Mexico founded the Seminario de Mineria, which was the first engineering school in all of America.

Although gold had sparked the exploration of our country, the first gold rush in what is now the United States was a dismal failure. An Italian, Marcos de Niza, was sent north in 1539 from Mexico City to search for the legendary "Seven Cities of Cibola" which were rumored to have streets paved with gold. De Niza and another Franciscan friar were accompanied by a Negro slave, Esteban. While de Niza delayed en route with his dying fellow priest, Esteban traveled ahead. As a Negro he was somewhat accepted by many tribes of superstitious Indians, but was killed at Zuñi pueblo. De Niza heard of Esteban's death as he approached Zuñi later, and decided not to enter the village. From a distance, the yellow rays of the setting sun made the village appear to be truly paved with gold. De Niza raised a cross to claim the region in the name of de Mendoza, the Viceroy of New Spain, then returned to Mexico with his story of the Golden City.

An expedition led in 1540 by Coronado was better equipped with cavalry and foot soldiers, and with several priests to convert the Indians to Christianity. Many of his troops were Indian auxiliaries from Mexico. This expedition apparently reached the Hopi towns in Arizona, where a group of the villages were known as Cibila, later corrupted to Cibola. Upon his return to Mexico, Coronado contradicted the report of Niza, but he did state that although the buildings were not golden, they were remarkably constructed of masonry without mortar.

The second gold rush in the United States was reasonably quiet, in a well settled part of the country. This region still produces a moderate quantity of gold, yet is far from the western mining camps usually considered to be the scenes of the gold rushes.

When Ponce de Leon landed in Florida in 1573 he was told of Indian tribes that had much gold. Later explorers reported gold ornaments made by the Indians from gold taken from streams in the southern Appalachians.

Actual mining began when a nugget "the size of a small smoothing iron" was found in 1799 at Reed, North Carolina. At this camp a lump weighing 28 pounds was later found. Thomas Jefferson told of a quartz gold specimen taken from the Rappahannock river. The U.S. Mint reported gold production from North Carolina beginning in 1793. From 1793 to 1828 the only reported gold production in the United States was from North Carolina, a total of only $110,000.00. In 1829 Virginia first reported production as a gold mining state. However, at the Belzoro mine, opened in 1832 a few miles from Richmond, ancient crucibles and other smelting equipment were discovered. Possibly there had been sophisticated Indian mining, or there had been unreported mining from some of the British colonies. The Belzoro continued production for many years, and following the Civil War was given wide publicity as the most promising source of mineral wealth in the country.

THE BELZORO GOLD MINE, IN GOOCHLAND COUNTY, VIRGINIA.

In 1829 there was a real gold rush to Georgia, followed in a few years by a strike in Alabama. The southern gold strikes resulted in United States Mints being established at Charlotte, North Carolina, Dahlonega, Georgia, and at New Orleans. Most of the workings were placer or open cuts until 1880 when some underground mining began. Placer mining, the form usually used in early development of gold regions, consisted of washing free gold from the country rock by use of pans, cradles, long toms, and other easily built equipment. These devices will be pictured and described later in this booklet.

From 1804 to 1866 the Appalachian production amounted to almost $20,000,000. Although this was a minor part of the Southern economy as a whole, it was very important in preparing the nation for a quicker development of the California diggings. The greatest part of the Appalachian production was between 1840 and 1849 when many miners went westward. These men were the experienced miners in California who taught the amateur goldseekers from Northern cities and farms how to mine gold. Later, after the end of the War, Southerners again gave men and skills to gold mining in the West.

It is obvious that gold was known in California long before the Sutter's Mill discovery which started the third American gold rush. Jedediah Smith had discovered gold in 1825 near Mono Lake while heading a party of trappers working for the American Fur Company. The next year Smith returned for further prospecting, but his entire party was killed by the Indians. The American Fur Company thereafter would not permit its employees to prospect for minerals.

Mexicans had mined gold near Los Angeles as early as 1842. However, prospecting was not encouraged for it took laborers from the fields, and it was feared that gold production would precipitate an invasion by Americans. The American Government, anticipating a change of ownership of the region, did not wish to call undue attention to its potential wealth. When the region did become American, and gold was publicly discovered, the nation benefitted immensely, not only from the hidden treasure, but from the worldwide stimulation of immigration, expanded international trade, and the new interest in exploration of wastelands on other continents.

In 1839 John Sutter, a Swiss immigrant, settled in Northern California. He was a resourceful, industrious pioneer who multiplied his holdings under a Russian land grand to create a small empire. When the Russians withdrew in 1841, Sutter bought virtually all of their holdings at Fort Ross, including ships, cattle, farm implements, and even an arsenal captured by the Russians during Napoleon's 1813 retreat from Moscow. He moved most of the materials and buildings to near the present site of Sacramento. His holdings included a flour mill, saw mill, tannery, and ranches feeding thousands of head of cattle. He was a leading fur trader, employing many Indians as trappers or ranch hands.

In January of 1848, James Marshall was completing a new saw mill on one of the Sutter ranches. Marshall was an able, imaginative mechanic, who had long believed that there was gold in the area. While testing the mill race for the water wheel, Marshall found a few flakes of gold. He excitedly collected about an ounce of gold, and dashed off on horseback to Fort Sutter, some forty miles away. Late in the evening he reached the fort in a driving rain. In the privacy of a closed room he announced his find to Sutter, who carefully tested the mineral to make sure it was not pyrite.

In spite of Sutter's invitation to stay with him, Marshall insisted on returning to the mill. Sutter came out the next day to find the men at the mill gathering gold, neglecting their work. At almost the same time the treaty of Guadalupe Hidalgo was being signed, ceding the area to the United States. However, the gold strike was unknown to either Mexico or the U.S. at the time, and the treaty did not take effect until the end of May. Sutter was alarmed at the possibility that the gold excitement would jeopardize the annexation, and that work would stop at his mills, tannery, and on the operations of his ranches. Therefore, he pledged his men to secrecy for six weeks. Within a few days, the news had spread, and Sutter's ruin began. His men left their work, his farms were neglected, and soon were vandalized by gold seekers. Prospectors dug where they pleased, stole his crops and cattle, ruined his fur trade with the Indians, and left Sutter to fight for his lands in the courts of law. He spent his entire fortune, yet his claims were not allowed. Finally, the State of California allowed him a small annuity which he drew for a few years until he died at Washington, D.C. in 1880.

JAS. W. MARSHALL, THE DISCOVERER OF GOLD IN CALIFORNIA.

Marshall was equally unfortunate. For a time he was moderately successful as a prospector, but he failed to hold his earnings. Both the discoverer and the original owner of the '49 gold were ruined by its discovery. In all probability a few thousand dollars are stilled owed to the Russian Crown for the unpaid balance on the Fort Ross purchase.

SUTTER'S MILL.

Of the 2000 men who had been in San Francisco in January of 1848, only five remained at the end of the year. The others, and three thousand more were at work in the mines. Average earnings the first year was $1000 in gold per man.

The miners of 1848 found gold so easily that they believed, and advertised, that there would be enough for any man who might come to the diggings. Two ounces a day was a common production for inexperienced miners. As gold was then $12.00 per ounce in cash, or $16.00 in trade, and as prices had not suffered serious inflation, work at the mines was profitable. Small claims sometimes produced as much as $100.00 per square foot. Nuggets of large size were occasionally found, but flakes and dust were much more common. Soil clinging to the roots of bushes gave as much as $50.00 when panned. There were the actual or imagined examples of the richness of the new land. There must have been many cases of gold taken from the grass roots, as it had been accumulating on gentle slopes for thousands of years. While lighter parts of the soil washed away, gold could be trapped by the plants.

Many of the early placer areas were opened by prominent ranchers who saw the methods, and learned to recognize gold areas, by visiting Sutter's. John Bidwell developed the placers at Bidwell's Bar on the Feather River. P.B. Reading recognized signs on his land, and opened diggings on the Trinity River. Many other ranchers also avoided the difficulties encountered by Sutter, who happened to farm the first, but not necessarily the richest regions.

In early 1848 there were two or three thousand Americans in all of California. By the end of the year there were 6,000. In July of 1849 there were about 15,000, and in December of 1849 there were 53,000. Within five years over 300,000 Americans had been drawn to California by gold. They came so fast that San Francisco had an ever changing population. Of the five hundred ships that entered the bay in the summer of '49, four hundred were deserted by passengers and crew, who left for the gold fields. Some ships were beached, becoming stores, hotels, or

Beached ships, San Francisco 1850

eating places. Over 20,000 men had come to California from Australia, China, Africa, and South America.

About half the Americans had come overland, and about half had come by sea, either rounding the Horn, or crossing the Isthmus of Panama on foot, or by pack train, and after 1855 by America's first trans-continental railroad, which speeded the journey between ships dropping passengers on the Atlantic side, and those taking on passengers to go up to San Francisco. The foreign immigrants were not as well educated as the Americans. Australians frequently were former convicts, while those from Latin America, and the Orient, or Africa were merely seeking better lives. A surprising proportion of the Americans, however, were doctors, lawyers, journalists, and former teachers. California had three or four times as many lawyers per capita as the rest of the country. The medical profession was stronger in California than in any other state, though many doctors mined, rather than practise medicine at first.

This sophisticated population developed a system of self government that became a model for the rest of the developing country. The early mining, land, and water laws were written of necessity, to make each mining camp work. The best of these finally were adopted by Congress, and still are in operation.

$2750 REWARD

WELLS, FARGO & CO.'S EXPRESS BOX

on LATROBE & FIDDLETOWN STAGE ROUTE, was ROBBED to-day, by two men, one mile from Plymouth, of $7000 in GOLD. We will pay

$500 Each

for ARREST and CONVICTION of the Robbers, and one-fourth of any portion of Treasure recovered.

JNO. J. VALENTINE, Gen. Supt.

San Francisco, July 5, 1875.

THE BAY OF SAN FRANCISCO.

Most of the early mining was placer mining which did not require heavy investments in equipment, and could be worked on a small scale by inexperienced men. Small companies were formed, to build dams, and diversion flumes or ditches, exposing rich ground under deep rivers.

Although many parts of California have produced gold, the Forty Niners mined in the area west of the Sierras. The placer mines were along the major rivers draining west from the Sierra Nevada. The ever present dream of the miners was to find the Mother lode from which the rivers were supposed to have taken the gold. When it was found, even greater riches would be developed.

In addition to permitting the use of primitive equipment, placer mining often resulted in the discovery of fantastically rich pockets of gold. This happened because the erosion of low grade ores over many years permitted concentrations to be made in the crevices of stream beds. The rolling action of the heavy, soft metal welded small particles together to form large ones, removed valueless gangue minerals from irregular crystals of gold, and forged them into smooth, durable lumps.

Except for minor pockets of "picture rock," most reliable lode gold mines are worked in large ore bodies of low yield. The reliability comes from the large reserves that will permit operation for many years. Once I was given a small piece of ore from the great Homestake Mine in South Dakota. This piece of ore had a tiny speck of visible gold on it. When I showed it to a miner he told me that it was the first time he had seen actual gold in Homestake ore during all his years of working there.

Courtesy of Knott's Berry Farm

Since the men who made the gold rushes were often amateurs of no experience, it was natural that they would succeed only in areas of obvious, visible wealth. They were not prepared to make the investments and be successful where it was necessary to mine ores that were unprepossessing in appearance, or ran only a few dollars per ton in values that were invisible.

BIG NUGGET OF GOLD.

THIS nugget, now on exhibition in San Francisco, weighing 95 pounds, and valued at $22,000, is the largest piece of gold ever found in California. It was taken out from the Monumental Quartz Mine, situated about twelve miles north of Downieville at the Serra Buttes. The nugget was found at a depth of 25 feet from the top on the 18th of last August. There had been taken from the mine up to the 1st of September $70,000 with the labor of three men; and it is still paying at the rate of $1500 per day, the gold being washed from the decomposed quartz with common pans. Our picture is from a sketch by L. R. TOWNSEND.

The first lode mine was discovered in a quartz vein near Mariposa, early in 1849 by Kit Carson. By July of that year the first stamp mill in the state worked the ores from the Mariposa. Because the first output of the mine was rich "picture rock" ore, early mining on the lode consisted of digging gopher holes for the high grade ores, and crushing them in the mill. There was no equipment for amalgamation, so the crushed ore from the mill was cleaned up in a rocker. Later an arrastre was built to handle the work. In June of 1850 rich gold quartz was found at Grass Valley.

SHEEP-CORRAL LODGINGS.

Prices climbed and became outlandish. Shir's were $25.00, pork $220.00 per barrel, sardines $35.00 a dozen, flour $ .75 a pound, and candles $3.00 each. Shovels were $10.00 with picks at $15.00 and miner's pans at $9.00. Women were scarce, and ladies even scarcer, which made Ralph Andrist's term, "The Big Stag Party" particularly appropriate. The men drank, fought, worked, and danced together in the isolated camps. In the cities the bars and gambling houses were ornate, and often better furnished than those in Eastern Cities. Fancy houses and red light districts began to grow in most of the camps. As wives and families were brought in, these districts were somewhat isolated from the family residential districts. Within a few years the major camps had become cities, with schools, churches, hotels, and some sort of official building. Masonic and other lodges provided social activities that helped dignify the new towns. Frequently the buildings were constructed from funds assembled by fines and court fees levied by the elected alcaldes or mayors of the towns.

The Law Office

Often the elected town governments failed to mete out justice fairly or swiftly. Committees were formed of Vigilantes, who held trials and carried out the executions quickly, and usually fairly. These vigilance committees, with the miners' meetings, formed many of the codified laws of the districts. Since there were no useful prisons, and few jails, quick justice was necessary. Punishment might be whipping or banishment for minor crimes, but usually death for crimes which would have drawn prison terms in the East. This is not as harsh as it sounds. When prisons were later established, they were miserable places with almost 100% mortality in the first few months.

A STATE OF SUSPENSE.

The Third Gold Rush into California was the one that helped create such literary figures as Old Block, Bret Harte, Mark Twain, and Joaquin Miller. These men, and others, wrote for the western papers and eastern publishers, describing the lives and works of the men. As California settled down, and other gold strikes were made, some of them, such as Harte and Clemens worked in other districts, among them the Washoe of Nevada.

From a pictorial standpoint, the early California Gold Rush was not well covered. Until 1851 Dagguerotypes were the only practical photographs made. These could not be reproduced easily, but were occasionally used by lithographers or etchers to prepare plates for illustrated books. A few lithographs and etchings have been reproduced in this booklet. After 1856 or so, wet plate processes and equipment were reliable enough to use in field photography. A few California pictures, and all of those on the later strikes were photographed by this method, which closely approximated present day photography in results. The California and Washoe rushes were not given complete journalistic coverage. Transportation was too difficult and too slow. However, many diaries were published, as well as illustrated guide books. These described life in the diggings, and some gave directions on how to succeed. Often the works were humorous; sometimes they were merely travel folders. Because illustrations were expensive, they were held to a minimum. Therefore, we find that there are more specific examples of mining practises and equipment taken from the Colorado workings. Since virtually all of the equipment used in Colorado had been tried in California, there is a reasonably complete pictorial coverage to the art of winning gold.

GOLD WASHING IN THE CALIFORNIA MINES.

As each mining district developed it became more civilized. The women came to make homes. Schools, churches, theaters, and hotels became prominent, and cities grew up where there had been camps of tents and cabins. There was usually plenty of money, freely spent, with the result that development was quicker than back in the East. Ice and snow refrigeration let Mother Lode hotels serve such delicacies as oysters, lobster, and other seafood as early as 1854. The bars served wines to tickle the palates of the most sophisticated, as well as to remind the South Americans, Australians, and orientals of home. Ornate bars would cater to dirty men fresh from the creeks, standing side by side with the city's well dressed financiers. The ladies of the evening shared their attentions with equal democracy, provided there was temporary financial equality, if not social prominence.

In spite of the backbreaking work, the sickness and hardships, and the scarcity of tools, the amateurs working from 1848 through 1852 managed to take out a quarter of a billion dollars in gold before the rich placers began to play out. By this time some tunnels had been driven into fossil placers, old stream beds covered by lava or other rocks. A few quartz lodes had been developed earlier, and the true "Gold Rush" of rich strikes was over. The hard rock mining that was necessary required expensive equipment, with large investments in shafts, tunnels, and hoists long before a profit could be realized. This was the field for professional promoters who hired many of the early miners who could not make their livings mining alone.

VIEW OF MARIPOSA

The glamor was gone, so when news came over the mountains that a new lode had been discovered in Nevada, men were ready to move eastward. The rich gold and silver mines of the Washoe had the same breed of miners, but a new kind of mine owner. Many mines were absentee owned. Many a prospector sold his mine to a syndicate, sometimes keeping his earnings and making a profit; sometimes selling much lower than he should. Henry Comstock himself received $11,000 for a mine that later produced the Hearst fortune, and Henry was the man who had made the Washoe district succeed.

The Nevada mines were in "bad Ground," subject to caving over excavations and tunnels. Here were developed some of the modern timbering methods still used in many mines. Square set timbering could give safe support to large stopes or mined-out underground areas. These even permitted the use of underground space for pumps, hoists, and other power equipment to make mining more efficient. Much of the equipment was operated by compressed air, pumped from a distance by water power. Compressed air operated drills, hoists, and even small locomotives for underground haulage. Here, and in the larger Colorado mining camps, mining came into its own as a precise scientific field. Washoe trained mining engineers. Colorado trained a new kind of ore dressers, who soon learned to operate mines on large tonnages of low grade ore. These developments permitted some mines to remain in operation long after rich ores were depleted. The Homestake Mine, America's largest producer, made a profit in 1880 on ore averaging $8.40 per ton. Greater tonnage of poorer ore is profitable there even now. The Black Hills are now contributing to the fund of mining knowledge begun when early miners from Georgia helped develop the California diggings.

OREGON MILL, UPPER AUSTIN.

DOWN THE SHAFT.

Staking a Claim on the Washoe

The Washoe cities were scenes of more speculation than had been seen in California. Newcomers to Virginia, Reno, and Silver City were besieged by stock touters, or by unemployed miners, each of whom was sure he was worth at least $50,000. There was very little new money in the camps, except that brought by newcomers, yet everyone was a millionaire on paper. By April of 1860 about two per cent of the inhabitants were miners. The rest were "investment advisors," bartenders, pimps, or prostitutes, with a few merchants and journalists. In spite of the heavy overload of non-productive residents, the mines did prosper, and maintained incomes for their owners although speculation got out of hand. The Washoe boom collapsed, but the mines continued to make money. More than three hundred million dollars came from the district before 1863. This helped finance the Union during the Civil War. Later there was another Washoe boom in the 1870's, which strengthened the Hearst fortune, and made that of John Mackay, a miner from the Yuba River. Mackay made a scientific study of mining, and later endowed the Nevada School of Mines. Using profits from the Comstock Lode he became a major organizer of American Telegraph & Telephone Co.

Most of the Southwest had enjoyed precious metal production to some extent before and during the California Gold Rush period.

In 1629 Geronimo Salmeron, a Franciscan monk, wrote a description of the many mineral deposits of New Mexico. His chief complaint was that the Spanish not only were too lazy to mine, but often destroyed the workings of men who did mine. Among the minerals he mentioned as being abundant were silver, copper, lead, sulphur, and the turquoise which the "pagan Indians regarded as precious as diamonds."

The first English account of any Southwestern mine was made by Zebulon Pike in 1806. He discussed the Santa Rita copper mine which is still being worked near Silver City, New Mexico. Twenty thousand mule loads of copper were produced each year even then.

DEEP MINING IN NEW MEXICO, 1680.

The Ortiz mine was opened near Santa Fe in 1833, and became a good producer of gold. It has often been named the oldest lode mine in America. However, the nearby Cerrillos district had the Mina del Tierra, which was opened near 1735, and worked for its silver and lead. There is evidence that the Jesuits had required Indians to work the mine as early as 1680. Ladders consisted of notched logs, running from one level or landing to another, and reaching as deep as 200 feet. The ladders were almost identical with ancient pueblo ladders. Stone tools were in use, and at one place the remains of a boat were found underground. Apparently water had flooded part of the workings, but formed an effective haulage system.

Colorado's mining industry was opened by men who had left the Appalachian mining districts to go to California in 1849. Dr. Levi Russell, with his brothers Green and Oliver had started West with four other Georgians and a herd of horses. In October they were warned against crossing the mountains, and made camp for the winter near the present site of Denver. They found gold in several locations along Cherry Creek, and saved the dust in goose quills. In the spring they continued to California where they mined near Downieville. They sold their California claims and returned to Georgia in 1857. Remembering the possibilities of Colorado gold, the entire Russell party with four new members returned to their Cherry Creek camp in August 1858. Early prospecting parties ranged as far North as Boulder Creek, and as far west as the Central City area where they found interesting prospects. It was too near winter to begin large scale operations, so part of the group remained in Colorado while six went east for provisions.

The Russell gold production for the summer of 1858 probably was less that $1000.00, but premature reports had multiplied it many times. The enthusiasm of Easterners was aroused again with hopes of a repetition of the California discoveries. During the winter of 1858 several hundred gold seekers were camped in what is now Denver. Many of the would-be miners were almost destitute. The towns of Auraria and Denver were too small to

REPRESENTATION OF MINERS WEIGHING THEIR GOLD.

absorb them. Food and shelter were inadequate. Flour was $40.00 per hundred pounds. Meat came from slaughtered draft animals, and from such game as was brought in by hunters and Indians. There was talk among some about lynching the Russell party. Most of the difficulties came from irresponsible reporting, and from embellished stories of how easy it was to get gold. One newspaper had written that Pike's Peak (almost 100 miles away from the placers) was a solid mass of gold. Writers had told of riding sleds equipped with blades down the slopes, planing up shavings of gold as they went. The feeling was at its highest during the cold winter. A few of the more experienced men left the towns to follow the rivers upstream to the mountains.

George A. Jackson, a cousin of Kit Carson, camped near what is now Idaho Springs. Early in January 1859 he found placer gold. By keeping a fire all night he managed to thaw the sands of a likely looking bar. After panning a half ounce or so of dust he secretly marked his location and returned to Denver. The silence he kept was unusual among gold seekers. After the spring thaws he returned with his partner, Tom Golden, and panned out several thousand dollars. Now the secret was much too big to keep. The result was a stampede up Clear Creek Canyon, led by the suddenly eager Denver "winter residents."

Pike's Peak

At about the same time John Gregory opened his placers at Gregory Gulch, and Green Russell began operations at the gulch bearing his name. The three major camps were located within a three mile radius but were on different stream drainages. Years later they were connected by a series of tunnels joining the lode mines developed in the same camps. The news spread rapidly and a flood of gold seekers began to cross the plains.

GREGORY GOLD DIGGINGS, COLORADO, MAY, 1859.

A major stampede for the mountains began in the spring of 1859. A hundred stage coaches were along the route from Leavenworth to Denver. Although many men went westward by coach, paying the one way fare of $100.00, most emigrants had their own transportation. The Republican River trail was crowded with wagons, coaches, pack trains, and men carrying their supplies on their own backs. Some used handcarts to take along more equipment and supplies. Many started too early in the season, and suffered from frozen hands and feet. Others carried too little in the way of provisions, so lived by the generosity of those who had enough. Some carried too much in the way of

MINERS' PROSPECTING.

luxuries, and littered the trails with discarded cookstoves, clothing, or tools. Some had to feed flour and other human provisions to their draft animals because they failed to find pasture. Others were lost trying to find short cuts to the mountains, and died of starvation or exposure.

In the midst of this stampede there were reports that the much advertised gold had been imported from California which caused a panic almost as contagious as the gold fever. Thousands of emigrants turned back after having gone less than a hundred miles west of Leavenworth. One party turned back when it was within twenty-five miles of the mountains. The first coach from Leavenworth completed its round trip back from Denver on the twenty-first of May, 1859. It carried the first shipment of gold dust, worth only $3500.00, but was escorted into Leavenworth for an official reception. From that time on the eastern metropolis of Kansas Territory gained stature with each report of new gold discoveries.

The rapid growth of the new region required some sort of laws governing ownership of claims. John Gregory was instrumental in framing the first mining laws of the region, laws that have gradually developed into the present mining laws. Originally the discovery claims on lodes were 200 feet, with 200 feet allocated to later settlers. Placer claims were 200 feet along the creek, from bank to bank. Although the actual dimensions were changed, the early mining and water laws helped the districts to grow rapidly. Because the wealth was divided fairly among many miners there was a free circulation of money. On the other hand, when mechanization did become necessary, there were no large mining firms to finance deep shafts, hoists, or mills. Many fortunes were established by early miners, and many were lost. Gregory was rich as a prospector, but died broke, an unsuccessful gambler, in 1864. One of the men who succeeded was George M. Pullman. After five years in Gregory Gulch he sold 2½ claims for $1000.00 per foot. The quarter of a million dollars started him on his industrial career. His idea for sleeping cars is said to have come from double deck miners' bunks used in the district.

Soon the little camps in the Gilpin country along Clear Creek became crowded. Parties of prospectors began crossing the mountains. In July of 1859 the first discoveries in South Park drew thousands of hopeful men to Tarryall Creek. The large claims left little for the late comers who then moved over to the South Platte, naming their town Fairplay because the small claims meant Fairplay for more late arriving miners. Rivalry between the early camps in South Park was stronger than in the mountain regions, possibly because the towns were more accessible to each other though more widely separated. Tarryall and Hamilton across the creek had bitter feuds. First Tarryall had a postoffice. Hamilton got a newspaper. Then Tarryall opened a mint. At the height of the feuds Fairplay proudly announced that she had a postoffice AND three newspapers. Many of the newcomers were from the Middle West. Jefferson, Alma, and Breckinridge were settled by men who brought the name of their state to Hoosier Pass, providing access to the mines on the Blue River.

North of the Denver settlements new mines opened in the Saint Vrain and Boulder Creek districts. Denver became a city with a newspaper when the Rocky Mountain News began publication on April 23, 1859. Before the end of the year fifty small stamp mills and mercury amalgamation plants were in operation. An express line provided regular service between the Missouri River and the Rocky Mountains. In 1860 Clark, Gruber & Co., opened a mint as part of their banking house. It operated until 1862 when the mint was purchased by the U.S. Government as an Assay office and gold refining smelter.

BANKING-HOUSE, DENVER CITY, COLORADO—MINERS BRINGING IN GOLD DUST.—[Sketched by Theodore R. Davis.]

Each decade seemed to produce a rush to another area. While Colorado was developing her gold and then silver mines, there were strikes in South Dakota which had a direct connection with Custer's death. The gold of the Black Hills was well publicized, but a writer in Scribner's Magazine "proved conclusively" in 1877 that the low grade ore, the shortness of the summer, and the tremendous distances for transporting supplies would prevent those mines from ever succeeding. Today the Homestake, opened near Deadwood, still produces gold, and has been the steadiest gold producing mine over the longest period in the entire nation. It opened in 1876, and has made a profit ever since.

LOWER END OF DEADWOOD.

In 1870 a thirteen year old boy located a claim at Julian, California, not far from San Diego, The claim there produced some $15,000,000 before the strike in Tombstone, Arizona in 1880 took the Julian miners on another merry chase. Tombstone became a successful camp, and because of the efforts of some of the non-working populace, earned a reputation as being the headquarters of the Wild West. Its mines became quite deep, encountered much water, and were among the first to require extensive pumping systems. As much as 8,000,000 gallons of water were being pumped daily from the great Contention Mine. An accidental stoppage of the pumps flooded lower workings containing equipment, and the mine could never be de-watered.

Courtesy of Knott's Berry Farm.

Gold strikes in Western Arizona and the Mohave region were accompanied by long lasting legends of lost mines and buried treasures. Only recently a man who prefers to remain anonymous has offered stories, samples, and pictures to support his claim that he has found the celebrated Lost Pegleg Mine, and has disposed of several hundred thousand dollars in gold in the past few years.

Leadville, Colorado, began as a gold camp, then produced silver steadily until the silver collapse in 1893. It continued as a mining camp until the end of World War I, producing lead, zinc, and manganese. It now is again prospering because of the huge Climax Molybdenum mine high on Climax Pass. At about the same time Colorado's mining fortunes again picked up with the growth of Cripple Creek, a new breed of mining camp, based on gold tellurides. These compounds had been ignored as they were not obviously gold. Cripple Creek mines went deep, required special metallurgy, and were the foundation of another group of Colorado fortunes, centered in Colorado Springs and Denver. Endowments of two of these fortunes still finance hospitals, colleges, and many other philanthropic activities.

VIEW IN CHESTNUT STREET, LEADVILLE

The last of our bonanzas was opened in 1896 along the Yukon River. Early Russian and American explorations had shown some gold, but no one was excited. Even Ed Shieffelin, the discover of Tombstone, found little of interest in a year's work along the river.

PLACER MINING ON THE KLONDIKE RIVER.

Diggings in the Arctic were difficult. Placer mining of the gravels required that they be thawed from the permafrost, and saved for warm weather for normal sluicing operations. Publicity on the Klondike and Yukon finds, both in Canada and Alaska, brought great waves of inexperienced men. Some represented small companies formed in the cities. Others came alone. Many died before they reached the gold fields. Travel usually was by ship to Haines, near Juneau, then on foot over Chilkoot Pass. A narrow trail led over the pass, and down to the string of lakes and creeks on the upper Yukon River. Often there were solid lines of men climbing the pass. Some could afford to hire Indian packers to carry their gear up the grueling climb. Others did it alone, carrying food, clothing, and equipment. The men who survived loved it and hated it. Their joys and miseries have been well described in the verse of Robert William Service. Today gold is produced in the area, by dredges and heavy placer equipment. A new boom in Alaska is promised not only by the oil fields being opened, but by the agriculture, tourism, and industries that seem to invade a wilderness faster if a gold rush has paved the way.

Fortunately, although the wildness, the glamor and romance have gone from the camps with their industrialization, their history remains. It is more comfortable for us to see them, and it is safer and more comfortable for those who live and work in them. Many do remain, or have been restored, to show to visitors the last days of our frontier history.

Most states where there have been gold rushes afford good opportunities for the tourist and visitor. The old camps helped new towns develop. Prospector's trails have become highways, so it is usually not difficult to visit many old mining camps. The buildings may be gone, or the camps may be renovated for tourists, but they are all interesting. They often are in beautiful areas.

The Mother Lode towns in California may be visited by driving State Highway 49, which winds along the old stage routes between mining camps, and gives an easy approach up the gulches to the isolated camps. It is a pleasant highway, rich in scenery and the history of the earliest land boom in the West. The town of Columbia is completely renovated as a state park. In Southern California at Knott's Berry Farm, Disneyland, and other parks, as well as in the desert region, one can find completely rebuilt mining camps. Equipment has been put into working order so demonstrations can be given to the visitors. In some locations the tourist can pan his own gold for a fee, usually with a guarantee of finding some color.

Virginia City, Montana, has a rebuilt ghost town, complete with all the buildings and businesses that were there during its heyday. There are others in Nevada, Utah, and Arizona, which can be found in your tourist guides. Oregon, Washington, and Alaska have them.

In Colorado, Central City and Blackhawk, a few miles from Denver, are almost unchanged. You may visit the old buildings, see the mines, or attend the Opera House plays in Central. Leadville, Fairplay, Cripple Creek, Silverton, Aspen, Ouray, and Telluride, as well as smaller towns are all interesting, and all are good examples of the mining camps of their particular periods. There are still operating mines in most of these towns, surrounded by some of the country's greatest mountain scenery.

DENVER CITY.

BLACK HAWK CITY.

CENTRAL CITY.

The earliest deposits of gold were alluvial gold, or placer gold. (Pronounced *plasser* by all but the rankest tenderfeet.) These were free, metallic gold concentrated by erosion because of the greater weight of gold compared with other minerals. Depending on its purity, gold may be from four to six times as heavy as ordinary sand or clay. It is heavier and more durable than most of the truly heavy minerals with which it is associated. In a stream, or on an eroding hillside, it moves more slowly than the less valuable minerals. Therefore it can be concentrated in two ways that were useful to the old prospectors. A short explanation of how ore deposits are formed will help show how the successful prospector located his claims.

Vein minerals were deposited long ago in cracks and fissures of rocks. Such fissures usually were formed during mountain building, so mineralized areas are more commonly found in mountainous areas. Hot liquids and sometimes hot vapors, under high pressure, were forced upward from the plastic semi-molten rock or magma underlying the earth's harder crust. These liquids and vapors were acid enough to dissolve many metals from the rocks through which they passed. As they were cooled upon contact with the fissure walls the less soluble metals or minerals were precipitated against the country rock. Sometimes the fissures were completely filled. In this way gold was crystallized out of the solution along with sulfides of various metals, such as pyrite (iron), galena (lead), argentite (silver), chalcocite and chalcopyrite (copper), and quartz (silicon dioxide). At times such deposition might occur quite near the surface, and at lower temperatures. Then the most volatile minerals would be deposited. The gold tellurides at Cripple Creek, silver ores at Tonopah, Nevada, and mercury ores of California were formed in this way. However, most of the gold deposits were formed at greater depth.

Probably similar vein deposits are being formed at present in many parts of the world. However the veins being mined today, and the veins discovered during our gold rushes were formed many thousands of years ago. They were buried under hundreds of feet of rock in our mountain areas to be exposed gradually by the decay and erosion of the overlying rocks.

For thousands of years the rock was broken into smaller particles by action of frost, wind, and rain. Plants turned it into soil by dissolving part of the minerals, rains washed the soil down hill, and eventually down stream. Finally, an ore deposit might be uncovered. Air and water dissolved out some of its minerals. Pyrite, which usually was present was oxidized to a porous, rusty mass, and also produced sulfuric acid which helped speed up the solution of the minerals. The light minerals were washed down hill. If the exposure was on an almost level location, or on a gentle slope, gold might remain near its original location, and would be concentrated by removal of its neighbor minerals. This gold would be found in rough, angular grains, closely resembling the original crystals. The rusty debris near it would often stain the outcropping quartz vein, forming a gossan, or iron hat, virtually a sign post for a likely place to prospect. Some of the gold might work its way down hill, leaving a faint trail by which a prospector *might* locate the "mother lode" of the placer he had been working.

Often a considerable amount of gold would be washed down hill to find its way into a permanent stream bed. Here it was separated from scraps of quartz by the milling of the stream. Quartz pebbles or boulders would be ground up, sometimes releasing more gold. The particles of gold gathered together where the velocity of the water was slower or where there were pockets. Some small particles were welded together by the pounding of thousands of gravel particles forming larger nuggets. Undoubtedly the reverse process also happened with huge crystals being reduced to small flakes. Whenever the speed of the stream was reduced, gravels would drop out of suspension. Gold, being much heavier than the gravel would collect in crevices and pockets in the bed of the stream.

Gold left by water on hillsides or in stream channels is called placer gold, and usually has a higher fineness or percentage of gold than vein gold. Placer gold is sometimes 900 fine (90% gold, 10% other metals) while vein gold is 500 to 800 fine. This is because silver, copper, and other metals are dissolved from the outer layers of the grains as they are being pounded and transported by the streams.

The qualities of malleability, insolubility, and weight, put the gold where it could be found. Its color made it noticeable, and all it qualities gave it its value to men from earliest times. Almost all of the gold that was ever mined is still in use. Through the ages gold has been the most important placer mineral.

If a miner located a lode from placer signs, he would find his gold relatively easy to mine and separate near the surface. However, as the depth increased he would find there had been little weathering to concentrate it. He might find another concentration near the upper surface of the water table, but the deeper he dug, the less he could act as a prospector, and the more he had to operate as a miner, with better metallurgical processes, and usually had to combine with other miners for the treatment of his ores. Thus the mining districts grew from simple operations to industrial complexes.

During the early stages of any district the mining was simple. Usually the men were not well equipped, and often the strike was made accidentally.

In the early placers near Los Angeles, the Mexicans used pocket knives. Marshall used a knife at Sutter's Mill to pry gold out of the crannies in his tail race. For want of better tools, the common knife was used to dig away at visible gold, and to loosen gravels for washing. The word *placer* comes from the Spanish word for sand bar or gravel bank. Placer deposits are those where the gold or other valuable metals have been moved by erosions from their place of origin, and deposited again. Because the valuable metals are heavier than ordinary sand and rock, they collect in lower pockets, or wherever the velocity of a stream might be slowed down. The heavy metals fall first, while lighter particles are carried on.

Because gold is heavier, it can be separated from other minerals by its weight. The early Mexican miners used a batea or shallow wooden pan. The Anglo miners coming into California used bateas, shovels, frying pans, or a special large metal pan which has changed little from that day to the present. The pan was filled about two thirds full of the gravel to be washed, and then dipped in water. The miner stirred the contents of the pan with his bare hands, breaking lumps of clay, and throwing out the larger pebbles. After all lumps were broken up, the pan was held in both hands, still under water. By a combined rocking and circular motion, the lighter sands were made to slide over the lip of the pan, while the heavier sands concentrated at the bottom of the pan. When all the light sands had been removed the pan was picked up, and by a whirling motion, the heavy sand caused to string out along the angle between side and bottom. There would be black magnetite or iron oxide, followed by a slower fan of "color" if there was gold in the pan. If there had been larger particles they were buried in the black sands.

Most of the early digging was done with pick and shovel. It was hard, backbreaking work, as many tons of dirt had to be dug and moved to wherever the gold was to be separated from the gangue minerals. If the digging were in a stream bed, the work was also cold and wet. Many miners in the California rushes could not survive the damp and cold of mining in the winter, even where the temperature rarely fell below freezing.

Wheelbarrows were sometimes used, but impatient gold seekers often would not spend time in making familiar equipment to make their work more efficient.

Usually the black sands and the gold were saved for a later cleanup, while more gravels were panned to produce a supply of the concentrates. The gold was later separated from the black sands by stirring gently with a magnet, which lifted the iron oxide. Care had to be taken so no gold was trapped between the particles of magnetic sand. It was sometimes removed by drying the concentrate, and gently blowing away the lighter black sand on a board, or in a shallow scoop. More commonly, the concentrate was treated with mercury which would wet the gold, and partly dissolve it. After being used for a time, the mercury was cleaned by filtering through a chamois skin. The gold-mercury paste or amalgam remaining was then heated to yield a button of pure gold. The mercury was distilled off and condensed to be re-used. Early in California's gold mining the area around San José produced mercury for the amalgamation process.

QUICKSILVER MACHINE, IN MORMON GULCH.

WINNOWING GOLD, NEAR CHINESE CAMP.

If no water was available, gold was concentrated by winnowing on a blanket. The dry ore was placed in a blanket held by two men. They would toss it up into the air so the lighter sands could be carried away by wind, much as chaff was removed from grain. Then the concentrates from this were treated to separate the gold. Blankets were hard work, and were used only in dry camps. Columbia, California, which became one of the more prosperous camps, had practically no water. After a system of ditches and tunnels had been dug to carry water to the camp, over $100,000,000 in gold was taken out. However, before water was brought to it, the camp had failed twice. It became known as the "Richest square mile on earth," a title later claimed by dozens of camps, including Central City, Cripple Creek, Dawson, and others.

The miner using the pan had to squat, often in cold running water. The pan of gravel and water is a heavy, tiring thing to manipulate, and on lower grade gravels, hundreds of pounds of sand daily required washing. Therefore various improvements were made. The first one-man improvement was the cradle. This consisted of a box built on rockers. It had a gently sloping bottom with cleats, and some sort of straining device at the top, either a perforated metal sieve, wire, or wooden slats. Gravel was loaded into the top, and as the device was rocked, water was poured over it.

The strainer let fine sand, and the gold wash down to the bottom. Pebbles, rocks and sometimes nuggets could be removed at the top. As the work progressed, more sand was added. The heavy concentrates collected behind the cleats, or in scraps of blanket or carpet tacked to the bottom. These were then cleaned in a pan and treated as any concentrates.

An improvement on the cradle was the Long Tom, which operated on a similar principle, but was rocked longitudinally on rollers made of logs. Two or more men worked it. One shoveled dirt in at the head, another would rock the 12 foot long trough, and a third carried water. Sluice boxes build along streams could use flowing water to do the same work. Gravel was shovelled into the box while a miner and the stream kept the material stirred up so sands were carried off and the heavy materials were stopped by cleats and blankets. Some sluices were a hundred feet long, worked by a small company of partners, and perhaps cleaned once a day or even less frequently. All the work was strenuous, wet, cold, dirty, and uncomfortable.

FLUTTER-WHEEL, ON THE TUOLUMNE.

When a group of men had found a promising location they might begin sluice operation, and build a dam to divert the stream completely, so they could dig clear to bedrock in relative comfort. Sometimes the stream was carried along in a flume much higher than the original stream bed. Water wheels in the flume operated pumps to dry out the workings, and to hoist

THE SLUICE.

gravels up to sluice boxes. Water from the flume fed the sluices. Shares in these partnerships were brought and sold, sometimes for as much as $2000.00. At Murderer's Bar in California, an extensive series of sluices, water wheels and workings were built, only to be destroyed by a flood from a cloudburst upstream.

Tremendous amounts of earth were moved by these primitive methods. Later Hydraulicking was employed. Flumes brought water down from higher elevations to feed through pipes and hoses to large nozzles called monitors. These were directed at the gravel banks to break them down and wash them into sluices for treatment. Millions of tons of sand and mud ruined good agricultural land and brought about strict legal controls over hydraulic mining in California. Hydraulic mining is still done in Alaska and some areas where it is permitted by law.

RIVER OPERATIONS AT MURDERER'S BAR.

With the advent of power equipment, dredging became useful in treating low grade place deposits. Boats carrying equipment for excavating and for washing the gravel, then for stacking the tailings floated in their own ponds of water in the lower rivers. A good dredge master could keep his pond full, and work over large areas by watching the flow of the stream, and the care with which he built his dam behind him. In 1965 a dredge began operation in Colorado on the Animas River above Durango. It had been built and launched in Benicia, California, and had worked off the coast of the Carolinas. It was then dismantled and taken to Montana where it worked a number of years. It was again taken apart, moved to Colorado, and reassembled and floated in a pond of a couple of acres. Now it is busily moving its pond along the stream. There are other dredges in Colorado now. One in South Park is awaiting better times for gold. One near Tiger has sunk in its pond. All that remains is part of the old wooden barge. Its dredge master had once worked up the stream so far above Tiger that he claimed he was awaiting a good spring runoff so he could go on over the Continental Divide. This is remarkable when one remembers that a dredge is really a mine and a mill built together on a boat.

Many of the fossil placers, where old stream beds had been covered by later deposits, or by volcanic lava or ash were mined by tunnels. The rock was stronger, and required milling to separate it from the ores, as did the quartz that engulfed the ores in lode mines. These required crushing or grinding to separate the gangue from the values. Sometimes this was done by hand tools, such as the pestle and mortar, but early in the mining period an old Mexican tool was put to work. The Arrastre was easy to make, and could be operated by animal power, or by water wheels. Its heavy parts were made from local stone, so freight was not a factor in the cost. Essentially an arrastre was a large rock dragged around a circular path.

The preliminary steps in making an arrastre consisted of finding a suitable rock surface, preferably where water could be run over it. A pivot hole was chipped in the center, and a roughly

circular trough was started by chipping. A frame of timber held a wooden pivot and arm, from which a large stone could be dragged. As the stone rotated it cut a doughnut shaped cavity in the base rock, and at the same time it ground the ore to a fine pulp. The light, brittle quartz was washed away, while the tough, heavy gold was not ground as finely, and remained in the trough. Often the arrastre was used to make the mixture of mercury and gold more intimate for good amalgamation. Some arrastres were refined to the point of having a pair of round stone rollers rotating in the trough to crush the ore. Throughout the Western mining regions one can still find the smooth circular cavities of arrastres that were built in stream beds where water could be used for milling.

Multiple mortars and pestles were built, operated first by water power, and later by steam or electricity. These stamp mills were an early development in the Mother Lode hard rock mines, and are still in use in many mining districts. The heavy pestles might weigh as much as a ton. They were lifted by cams on strong rotating shafts, and then were dropped onto the rock fed to the batteries of stamps. The finely ground ore was carried away to concentration equipment by a steady stream of water. Gravity concentration was often used with mercury amalgamation. Sometimes cyanide was used to dissolve the gold, which was then precipitated out by powdered zinc, and dried by removing the water from the gold sludge in a filter press. Sometimes a combination of all these methods was used, with cyanide removing the last gold from the tailings of the gravity processes. In later years flotation agents, which cause foam to adhere selectively to the desired metals, were used for a higher efficiency in the winning of precious metals. Improvements as time went on made it profitable to rework the tailings dumps of many early mines. Occasionally the dumps yielded greater profits than were taken from the original ore.

The hand methods of mining were used not only on placers, but in early underground lode mining. "Rat holes" were dug by hand, using picks, hand held drills, and by men crawling through small passages to scratch out the batter values. The methods used often were identical with those used for a thousand years in Europe. The structure of an ore body showed them the way.

Usually a lode was first located by following signs of placer gold up a creek, then up a hillside to where a gossan was found. The gossan was the weathered, oxidized outcrop of a group of gold bearing minerals. The gold might originally have been dispersed in pyrite or fools gold which filled crevices in quartz. At the surface the pyrite, a sulfide of iron was weathered into a rusty oxide by the action of rain, air, and sunlight. It often would swell, and crumble the surrounding quartz. Rain would wash away the lighter quartz, and dissolve some of the iron compounds, leaving concentrations of the gold in place. This was the gossan, "Iron Hat," or Eisenhut of the European miners. Iron Hat Science, or Eisenhüttenwesen, is still the German name for Metallurgy. Brown stains in white quartz became known indicators of gold, and were easy to dig. If a prospect was good, the miner dug deeper, following only the veins which he hoped would be productive. Before explosives, rock was broken by heating and dashing cold water on the surface, and by pecking away with hand tools. There were many such mines in Europe, and in early America. Now hand mining is done by drilling holes, and inserting charges of dynamite to break up the rock. Hand drilling contests were an important part of celebrations in our early mining camps.

Now drilling in almost all mines is done by compressed air drills, and in large mines by Jumbos, or multiple drills mounted on large machines to place the drill holes for most economical rock breaking. Broken ore is now carried to mills where it is crushed, screened, and ground to the proper size for the concentration process to be used for that particular ore. Sometimes it is treated as a fine sand. Sometimes it is a true mud of very fine particles. In a very few mines, the entire present production is by treatment of water draining from them.

Mechanization has now replaced human labor in almost all processes of metal production, from digging the ore to melting the refined metal. Our newest American gold mine is one of the most automated in the world.

The Carlin Gold mine in Nevada is operated as an open pit mine. About nine tons of overburden are removed to produce one ton of ore, which in turn yields about $10.00 in gold. The gold is invisible except at 1,500 time magnification, yet reserves are blocked out for production of $100,000,000. Earth-moving equipment removes the overburden and hauls ore to the mill. Here it is ground, leached in cyanide tanks, and the gold is precipitated from the clear cyanide solution by zinc dust. The mill is almost automatic, with a working staff of three men per shift. The Carlin mine, the Homestake, and Knob Hill in Washington, are the only gold mines of consequence now operating in the United States. All our other gold production is a by-product of copper and other metals. With the development of other automated low grade mines such as the Carlin, perhaps our production of gold can again meet our industrial needs.

In spite of the changes in mining, which make modern mines resemble city factories more than their romantic ancestors, some of the old mines and mining camps remain. They can be seen in almost any Western state. Rock hounds and "sun glass" bottle collectors can prowl the old mine dumps, or the old town dumps, searching for colorful minerals, purple glass, and antiques from the camps. Even if we can't always find gold, we can enjoy the scenery and have healthful exercise in good fresh air. If we visit the camps at the right season, we will see the biggest display of natural gold of them all. In the West aspens are the first trees to come back to devastated areas after fires, floods, or destruction by man. Aspens, with their clean white trunks, and quivering green leaves, are beautiful in the summer. In the fall the leaves turn to bright yellow, and finally make a golden carpet, covering dumps, cabins, and small deserted villages. Each fall sees another gold rush when hundreds of us go up into the mountains to view the new golden treasure.

The End